A Helpful Primer for
High School or College Chemistry

Illustrations by Cris Qualiana

ISBN: 0615386601
ISBN-13: 9780615386607

Table of Contents

TABLE OF CONTENTS...**III**

PREFACE...**V**

INTRODUCTION–THE BIG PICTURE...**VII**

CHAPTER ONE—CHEMISTRY HOMEWORK AND TEST-TAKING TIPS .... **I**

CHAPTER TWO—THE ATOM AND THE PERIODIC TABLE..................................**7**

    The Ancient Greeks 9

    Scale 10

    Models of Atoms and Molecules 10

    The Periodic Table... 13

CHAPTER THREE—SCIENTIFIC NOTATION **19**

CHAPTER FOUR—USING YOUR CALCULATOR....................................**23**

    Very Large Numbers... 24

    Very Small Numbers... 25

CHAPTER FIVE—SIGNIFICANT FIGURES **29**

    Significant Figures in Addition and Subtraction 31

    Significant Figures in Multiplication and Division 32

CHAPTER SIX—THE MOLE AND MOLARITY **35**

    The Mole... 35

    Molarity 37

CHAPTER SEVEN—UNITS AND CONVERSIONS **41**

    Units... 41

    Conversions 42

CHAPTER EIGHT—ACIDITY AND BASICITY...**49**

CHAPTER NINE—THE IDEAL GAS LAW **53**

    Solving Ideal Gas Problems.. 54

    Pressure and Volume ($PV = nRT$) 55

    Temperature (PV = nRT)... 57

    Number of Moles (PV = nRT).. 57

CHAPTER TEN—SOLUBILITY **61**

CHAPTER ELEVEN–BONDING..**65**

    The Octet Rule 66

    Bonds... 68

    Ionic Bond 69

    Covalent Bond.. 71

    Hydrogen Bond.. 72

    Metallic Bond 72

CHAPTER TWELVE—CHEMICAL REACTIONS **75**

AFTERWORD **81**

    How to Contact the Author 81

APPENDIX **83**

    Further Recommended Reading:... 83

ABBREVIATIONS...**85**

GLOSSARY **87**

INDEX **95**

Preface

I wrote this book because whenever I mention that I am a chemist to someone I've just met, they almost always say something along the lines of, "Oh wow, good for you, I hated chemistry!" This really bothers me, not because I feel the need to defend my profession, but because I think it's a shame that so many people feel this way about chemistry and science in general. Science has so many applications in everyday life, and learning it should be interesting and relevant. I don't believe that science should be a mysterious society that only certain people understand; even if you are only interested in getting by in chemistry and never taking it again, you should still get something useful from it. Plus, if you decide later that you are interested in pursuing science further, it is helpful to have a solid knowledge of the basic concepts. I believe that everyone should learn at least the basics of chemistry the way everyone learns to add and subtract.

When I was in seventh grade, I wanted to learn to play the flute. I signed up for band class and had one-on-one lessons with the school's band director. But no matter how hard I tried, I just couldn't figure out how to make music with it. The best I could manage was a sort of airy, tweeting sound. After a few weeks of trying in vain to teach me how to play, my music teacher asked me to buy a book called *Breeze-Easy Method for Flute*. For most kids who had some talent, this book might have been an insult because it was so basic. For me though, the way it broke down and explained everything in detail really helped. The book had good pictures that showed how I should hold my lips on the flute and exactly how much space there should be between my lips when I blew into the instrument. I was suddenly able to learn how to play, even though I really had little talent and was just trying to learn the basics, get through the class and maybe gain some music appreciation.

Many of you may have had the same experience I had with the flute in science classes. So much information is piled on so quickly that the concepts elude you. So you resort to memorizing equations and facts, hoping to get by with enough knowledge to pass the class. This can work in elementary school to some extent, but usually by the time you encounter high school biology and chemistry, memorization doesn't work anymore because of the greater depth of learning required. It becomes very important to do the problems and understand the concepts. Learning and then applying concepts is emphasized in high school and college, and if you don't work hard, it shows up in your grades. Unfortunately, this may also result in a lifelong dislike and distrust of anything scientific, which does you a disservice.

This book is written primarily for students who have had trouble with science in the past and are now about to take a chemistry or physical science class. It is written as a non-threatening introduction to the subject, ideally read over the summer so that you will already understand some of the concepts before you start the class. This book can also, of course, be used as a reference throughout the year.

I also think, based on my experience with adults, that many people forget most of what they learned in high school science classes, which is the last place that most people experience scientific training, unless they go on to study science in college. Often I think that people do not see how chemistry, biology and physics are all around them in their everyday life. I have tried to relate the chemistry concepts as much as possible to things encountered in everyday life so that they are easier to remember.

Please let me know whether this book helped you, after you have taken the formal course. Contact information can be found in the appendix. Enjoy!

Introduction

The Big Picture

"When am I ever going to use this again?"

I have to admit, you may never need to figure out how to produce a chemical compound. But it's pretty likely that someday you will need *something* from what you learn in chemistry. If you learn the concepts, it can at least make life more interesting, since you will know more about your environment.

Let's start with chemistry as a whole. What are we trying to do with chemistry? Well, first of all, we are trying to understand the world around us, the things we look at, touch and use—what are they made from? How do they interact with each other? How does your body work?

Secondly, with chemistry we can use the things around us to make our lives better. Take the example of plastic. There wasn't any plastic a hundred years ago: no cheap, light, strong material that could be molded into any shape. Carving stone and wood is time-consuming, and other materials like clay or glass are both fragile and heavy. Without the science of chemistry, we wouldn't have the opportunity to make newer, better materials like plastics that make life easier.

Think about the plastic tubing used at a hospital, artificial limbs, safety-sealed aspirin bottles and video game controllers. They were all created using polymers designed for these specific purposes.

Or how about deodorant—think about how unpleasant sitting on a crowded airplane or bus would be if there were no antiperspirant or scented soap. These products are all created using chemical reactions.

In order to study chemistry (or any science), it is necessary to learn what others have done before and build upon that, so that you aren't starting from scratch. Here are the basic things that you will learn about in high school chemistry:

- How to measure things correctly (significant figures, moles, units)
- What makes up the world around us and how it behaves (the periodic table, bonding)
- What happens when you react one chemical with another (moles, bonding, acids, bases)
- How these tools are applied (chemical reactions)

These tools are useful for understanding and using chemistry. The concepts can be overwhelming if you have no prior experience with them, which is why this book is designed to be read the summer before you will be taking chemistry. Sometimes having more time to allow concepts to "sink in" can really be helpful to your grades.

I'm hoping you have a different experience than most students (including myself), who learn each concept separately, do the problems and take the tests without seeing the big picture until years later, if at all. For me, this didn't happen until I was in graduate school.

Below is a diagram of some of the concepts you will learn in this book and how they fit together. This diagram is repeated at the beginning of most of the chapters in order to help you keep it in the front of your mind.

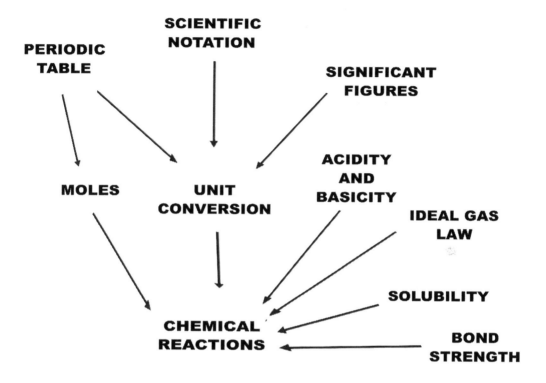

Here is an explanation of how each part is useful:

- **Periodic Table**—used as a tool to look up the properties of elements such as numbers of electrons and molar weights.
- **Scientific Notation**—used as a tool to help deal with very large and very small numbers.
- **Significant Figures**—used to make sure your measurements keep their original level of accuracy.
- **Moles**—help you collect the enormous number of atoms and molecules involved in a typical reaction into units that are more easily manipulated.
- **Unit Conversions**—a tool to help you convert from one unit to another and figure out how much product you can get from a reaction.

- **Acidity and Basicity**—knowing these properties gives you information about a molecule that helps predict the result of a reaction.
- **Ideal Gas Law**—allows you to calculate reactant and product amounts when one or both are in gaseous form.
- **Solubility**—helps you choose the right solvent for your reaction.
- **Bond Strength**—gives you an idea of how much energy is released when a reaction proceeds.
- **Chemical Reactions**—this concept brings everything together. If we are performing a reaction, we are using chemistry to create new things. If we are studying a reaction that occurs in nature, we can better understand the world around us, or our own bodies.

Putting all of the above pieces together:

1. Bond strength and acidity/basicity allow you to determine whether a planned chemical reaction is actually going to react.
2. Once you decide to perform a reaction, solubility information allows you to choose the right solvent so everything dissolves.
3. The periodic table, ideal gas law, moles and unit conversion are used to calculate how much of each reactant you need to measure out in order to get the right amount of product.
4. While you are doing your calculations, scientific notation and significant figures help you to get an accurate answer.

It is my hope that if a student who is planning to take chemistry or physical science reads this book, he or she will learn some of the important concepts up front and will then have an easier time as the concepts and equations are piled on during the school year. You have already taken a good first step by finding this book—I hope you keep reading.

Chapter One

Chemistry Homework and Test-taking Tips

MCHUMOR.com by T. McCracken

"There's really no need for confusion.
Part 95 of section 33 of article Q
in the formula quite clearly states ... "

I'm not going to give you any problems to do in this book, except chapter review questions, because you can find plenty of problems to solve in text-books or from your teacher. This book is just meant to be an introduction to basic concepts, and I don't want to bog you down. When you're done with this book, there are a few books in the appendix if you feel the need for additional preparation. What I want to do here is give you some help with doing problems in general (much of this is good advice for other subjects, too, not just chemistry).

Here is some advice that I didn't get until long after I could have really used it. When you're doing a chemistry problem, the best way to approach it is to make a valiant attempt at solving the problem and *try to get an answer of some kind, even if you are absolutely sure that it is the wrong answer*. After you have listened to class lectures and reviewed the information in your textbook, try the assigned problems. If your teacher gives you the answers ahead of time, then put the answer key on the other side of the room (or in the next room if you can't resist temptation), and don't look at the correct answer until you have written down an answer in your notebook that you have figured out for yourself. Don't even look at the textbook for hints yet!

Sometimes this involves pushing through a feeling of frustration and even despair, when you feel like saying, "I don't even know where to start!" This is normal; it's okay to just guess at the equation or process, write down some numbers and get a result. Get used to getting things wrong on paper when you do your homework, and leave room on the page for re-working the problem the right way after you see the correct answer.

After you have *your* answer, you can go look at the correct answer or look in your textbook for help if you don't have the answers in front of you. See if you got it right (great!) or figure out where you went wrong. Obviously, if you can't figure out what went wrong, you should review the material with your teacher until you understand it better.

I don't know why this works, but my students have always told me that it does. I think it has something to do with remembering mistakes more than remembering what you got right. It may be time-consuming, but it is worth it when you get to the exam and are better able to tackle those problems because you are prepared. You will have already found all the pitfalls in the calculations and identified your weak spots with the math. Many times you might understand the chemistry perfectly but make small mistakes in the math that cause you to lose points, especially if you get a teacher who doesn't give partial credit.

Another tip that might help you is to look up terms in the index of your chemistry textbook or on the Internet if you're having trouble understanding something in the book. It sounds obvious, but many people do not do this and just keep getting more and more confused as they do the assigned reading. I suggest you also use the glossary and index in this book as necessary.

Let me re-emphasize that you have to *do* the problems to get a good grade in chemistry. It may be tempting to think that if you just memorize the concepts you will be okay, but this is not the road to good grades. If you run out of problems to do, I'm sure your teacher would be happy to assign you more of them.

If the teacher provides you with sample tests, even better! Take the sample test in one sitting, a few days before the real test. Give yourself the same amount of time you will have in class to take the test, and don't refer to any notes, just as if it were the real thing. When you're done, take a break and then grade the test. This exercise will show you which concepts you still need to work on, and where your strengths lie. When you take the real test it's usually best to start with the questions you are most confident you'll get correct, then spend the rest of your time on the ones you had to review after taking the sample test.

Don't blame your teacher for your grades if you don't do the homework. His or her job is to present the concepts and teach you how to do the problems. Your job is to practice the problems so the concepts are reinforced in your head. You both have to do your jobs in order for you to get good grades. Many chemistry problems are tricky and you have to get used to dealing with the common pitfalls, like simple math mistakes and positive/negative sign confusion. Practicing by doing the problems is the best way to ensure that you catch these problems early and are able to fix them before the test.

There's a great book by Adam Robinson called *What Smart Students Know: Maximum Grades. Optimum Learning. Minimum Time.* I highly recommend this book, especially if you tend to get nervous about tests or sometimes freeze up during an exam. Many very intelligent students experience this—they study very hard but don't get good grades because of nervousness and anxiety. This book gives you strategies about organizing your time during a test, rehearsing for tests, etc. Again, use the problems in your school's text or in the books recommended in the appendix if you want to practice these techniques.

The first two chapters of this book may be a review for some of you, so you can skip them if you feel confident enough about atoms, molecules and scientific notation (this may be especially true if you have already taken physical science). I don't want you to get bored. However, I suggest you glance through these sections anyway as a refresher, or in case there's something there that you haven't seen before.

Chapter One Review Questions

1. What is the best way to do chemistry homework problems when the answer is given to you? Why do you think this method works?

_____

_____

_____

_____

_____

_____

_____

_____

_____

2. Where are two places you can look up the answers if you don't understand something in your chemistry class or homework?

_____

_____

_____

_____

_____

_____

_____

Chapter Two

The Atom and the Periodic Table

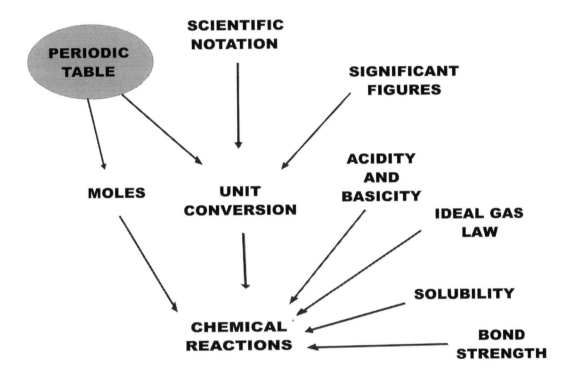

The basic reason for the study of chemistry is to figure out what's going on with matter, meaning the stuff all around us. That means the chair you're sitting in, for example. What is it made from? If you were building a chair, how would you know if the material it was constructed from would support your weight?

Before chemistry, this was pretty much done using trial and error—if you made a chair and it broke when someone sat in it, you would make the next chair out of something different and then keep trying new materials until you figured out the best one. Then you passed on that knowledge to your children, or your apprentices, and they continued making chairs out of wood or metal or whatever. Is your chair plastic? Plastic has been around for less than a hundred years. Where did it come from? How did we figure out how to make it?

Chemistry looks at things on an *atomic[1]* or *molecular[2]* level, so we can determine the basic properties of a material and then figure out how to do things at a larger level. We call this looking at things at the *micro* level. My goal is to get you to visualize the world around you on a micro level. This way of thinking may be a new experience for some of you, and is very valuable when learning chemistry.

Take a look at the chair you are sitting on. It is made of molecules, and if it's plastic or wood, these molecules are called polymers. At a micro level, polymers are like small spaghetti strands (*really* small, as I'll explain in a moment) that are stuck together so they form solid matter.

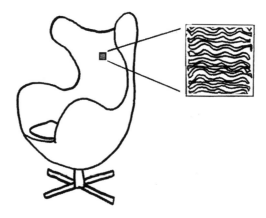

The other thing to remember about atoms and molecules is that they are always moving. Even the molecules that make up solids vibrate constantly. Molecules also contain energy in their bonds, which is what allows them to react

---

1   Atom—the smallest nondividable piece of something that still has the same properties of that thing.

2   Molecule—two or more atoms bonded together very strongly to form something with consistent properties.

with other molecules. One of the jobs that chemists have to do is measure molecular energy, so they can predict whether a planned chemical reaction will proceed or not, and how much heat will be required or given off by that reaction.

Think about a glass of water. Liquid water is made up of molecules of water that are only loosely stuck together, compared with wood or plastic. That's why water keeps moving around when it is sloshed or heated up, and why it's so easy to split into two smaller glasses of water. If you think of a glass of water at the micro level, you are clearly able to see why, from a micro standpoint, it doesn't make any sense to make a chair out of water at room temperature.

Now if the water were frozen, then you could make a chair out of it because the molecules would slow down, line up, and form a solid. Moreover, if you heat up a plastic chair enough, you can break the bonds that hold the "spaghetti" of polymer molecules together and the chair's material becomes a liquid. The temperature of the surrounding environment is important to consider when studying the chemistry of different materials.

THE ANCIENT GREEKS

About 2,500 years ago, two Greek philosophers, Leucippus and Democritus, came up with a theory that if you chopped something up into very, very small pieces, you would eventually get a piece of that object so small that you would no longer be able to divide it any further.

Here's an exercise you can easily try at home. Take a small piece of reg-

ular aluminum foil (about the size of a postage stamp) and tear or snip it into smaller and smaller pieces. You eventually get to the point where you just can't tear the foil up any smaller with the tools you have. Aluminum foil is mostly made up of aluminum atoms, with a few aluminum oxide atoms at the surface. However, for our purposes we will assume that it is made of pure aluminum. Aluminum metal is an element, which means it is a collection of only one type of atom.

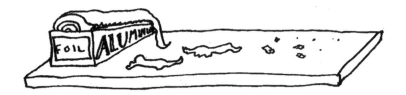

The Greeks thought that maybe a piece this small, or maybe a bit smaller, could be what they called the atom, defined as the smallest piece that still could be called aluminum. This made sense to them because in the absence of other evidence, people usually tend to think of the world around them in terms of what they can see. Of course, the ancient Greeks did not have aluminum foil, but the concept is still the same. They thought there might be atoms of wood, stone, fire, water, etc.

They also thought that those atoms might look like the substance they were made from. So water atoms were assumed to be smooth, soft and round, wood atoms hard, and fire atoms sharp and pointy (since it hurts to touch fire).

SCALE

The Greeks had no way of estimating the scale of atoms except what they could see with their naked eyes. They had no microscopes or sophisticated detection equipment. They probably assumed that the smallest piece that they could see was also the smallest possible piece of the material. It turns out that that tiny piece of aluminum that you have to squint to see actually has about 45,000,000,000,000,000,000 (45 quintillion) atoms of aluminum. How did I figure that out? We'll talk about that later, when we discuss the mole. For now, just look at that little piece of aluminum and think about just how tiny the atom is.

MODELS OF ATOMS AND MOLECULES

Because atoms are definitely too small to see with the naked eye, we need special instruments to detect groups of atoms and record data, so that we can analyze and draw conclusions about individual atoms and molecules. We have developed instruments that look at evidence of atoms' and molecules' composition and behavior, so we can figure out what they are doing and what they might look like. Then we develop theories about what they look like and try to go back again and verify those theories with more experiments. This is what science is all about. Below is an example of a mass spectrogram, which shows a molecule (ethylbenzene) and the breakdown products produced when it is hit with a beam of electrons. These breakdown products are used to determine which molecules are present in an unknown sample.

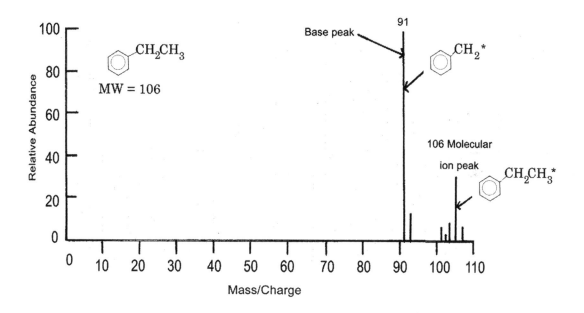

Atoms are made up of subatomic particles: a nucleus of protons and neutrons stuck together, surrounded by a cloud of electrons. All individual atoms look pretty much the same. Because we aren't able to see atoms or even molecules with our eyes, we use models to represent them and make it easier to imagine them. You may have already learned the Bohr model, shown below, which looks like planets orbiting the sun.

This is one way to show the atom, but it is not completely accurate because electrons do *not* actually orbit the nucleus like planets. Right now the best model we have for the hydrogen atom, for example, looks something like this, with an electron cloud around a nucleus:

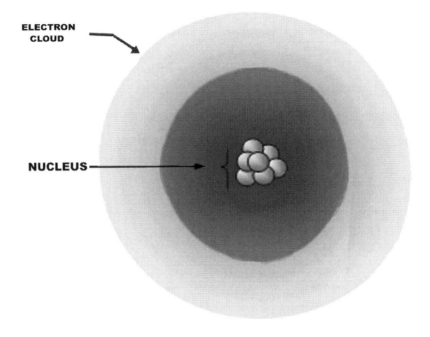

The hydrogen's electron is moving so fast that it would look like a cloud, if we could see it up close.

Once we start sticking atoms together to form molecules, they are going to have different shapes, as you can see in the figure below, a blobby, space-filling model of the phenol molecule:

In this model, the dark gray parts of the phenol molecule represent carbon (C) atoms, the black circle an oxygen (O) atom and the white circles hydrogen (H) atoms.

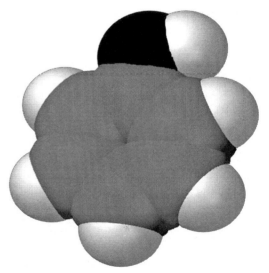

Chemists generally simplify this model more, to look like the following structure:

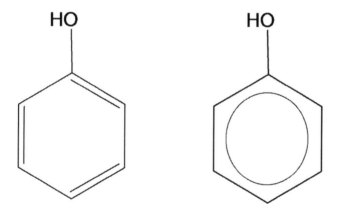

Organic chemists (who specialize in molecules containing carbon) may simplify it even further, to look one of these:

The last three pictures are just forms of shorthand for molecules. The simplest form is the molecular formula, which in this case would be C_6H_5OH. It's helpful to get used to visualizing the atom the way it really looks, so that you automatically picture the real molecule when you are looking at the symbols.

THE PERIODIC TABLE

The periodic table is a tool that is used to list the properties of atoms in an orderly fashion, so that you can look up atomic data quickly. Think of it as a database or a dictionary for atoms. The following page shows an example of a typical periodic table:

Legend:
- 1 — Number of protons
- H — Atomic Symbol
- Hydrogen 1.0079 — Number of grams per mole

1 H Hydrogen 1.0079																	2 He Helium 4.0026
3 Li Lithium 6.941	4 Be Beryllium 9.0122											5 B Boron 10.881	6 C Carbon 12.0107	7 N Nitrogen 14.0067	8 O Oxygen 15.9994	9 F Fluorine 18.9984	10 Ne Neon 20.1797
11 Na Sodium 22.9897	12 Mg Magnesium											13 Al Aluminum 26.9815	14 Si Silicon 28.0855	15 P Phosphorus 30.9738	16 S Sulfur 32.065	17 Cl Chlorine 35.453	18 Ar Argon 39.948
19 K Potassium 39.098	20 Ca Calcium 40.078	21 Sc Scandium 44.9559	22 Ti Titanium 47.867	23 V Vanadium 50.9415	24 Cr Chromium 51.9961	25 Mn Manganese 54.938	26 Fe Iron 55.845	27 Co Cobalt 58.9332	28 Ni Nickel 58.6934	29 Cu Copper 63.546	30 Zn Zinc 65.409	31 Ga Gallium 69.723	32 Ge Germanium 72.64	33 As Arsenic 74.9216	34 Se Selenium 78.96	35 Br Bromine 79.904	36 Kr Krypton 83.798
37 Rb Rubidium 85.4678	38 Sr Strontium 87.62	39 Y Yttrium 88.9059	40 Zr Zirconium 91.224	41 Nb Niobium 92.9064	42 Mo Molybdenum 95.94	43 Tc Technetium (98)	44 Ru Ruthenium 101.07	45 Rh Rhodium 102.9055	46 Pd Palladium 106.42	47 Ag Silver 107.8682	48 Cd Cadmium 112.411	49 In Indium 114.818	50 Sn Tin 118.71	51 Sb Antimony 121.76	52 Te Tellurium 127.6	53 I Iodine 126.9045	54 Xe Xenon 131.293
55 Cs Cesium 132.9055	56 Ba Barium 137.327	57 La Lanthanum 138.9055	72 Hf Hafnium 178.49	73 Ta Tantalum 180.9479	74 W Tungsten 183.84	75 Re Rhenium 186.207	76 Os Osmium 190.23	77 Ir Iridium 192.217	78 Pt Platinum 195.078	79 Au Gold 196.9665	80 Hg Mercury 200.59	81 Tl Thallium 204.3833	82 Pb Lead 207.2	83 Bi Bismuth 208.9804	84 Po Polonium (209)	85 At Astatine (210)	86 Rn Radon (222)
87 Fr Francium (223)	88 Ra Radium (226)	89 Ac Actinium 227.03	104 Rf Rutherfordium (261)	105 Db Dubnium (262)	106 Sg Seaborgium (266)	107 Bh Bohrium (264)	108 Hs Hassium (277)	109 Mt Meitnerium (268)	110 Ds Darmstadtium (271)	111 Rg Roentgenium (272)	112 Uub Ununbium (277)						

58 Ce Cenium 140.116	59 Pr Praseodymium 140.9077	60 Nd Neodymium 144.24	61 Pm Promethium (145)	62 Sm Samarium 150.36	63 Eu Europium 151.964	64 Gd Gadolinium 157.25	65 Tb Terbium 158.9253	66 Dy Dysprosium 162.5	67 Ho Holmium 164.9303	68 Er Erbium 167.259	69 Tm Thulium 168.9342	70 Yb Ytterbium 173.04	71 Lu Lutetium 174.967
90 Th Thorium 232.0381	91 Pa Protactinium 231.0359	92 U Uranium 238.0289	93 Np Neptunium (237)	94 Pu Plutonium (244)	95 Am Americium (243)	96 Cm Curium (247)	97 Bk Berkelium (247)	98 Cf Californium (251)	99 Es Einsteinium (252)	100 Fm Fermium (257)	101 Md Mendelevium (258)	102 No Nobelium (259)	103 Lr Lawrencium (262)

Notice that each atom's number of protons and electrons increases by one as you go across each row of the table (the number of protons always equals the number of electrons for a neutral atom). In fact, all of the atoms could just be listed in one long string, but putting them into a table like this allows us to identify trends.

Each column in the periodic table is called a *group*, and each row is called a *period*. Chemists often talk about atomic properties that increase or decrease "across a period", which simply means from left to right across one of the rows. Also notice that there are two long rows inserted in between barium and lutetium (Ba and Lu), and between radium and lawrencium (Ra and Lr). These are called the lanthanides and the actinides. You won't be doing much with these in high school or even introductory college chemistry, but you should be aware of the way the numbering changes in that part of the table. I suggest that you print out a copy of a periodic table from the Internet with the names of the atoms included for reference.

The three main pieces of information contained for each atom by the periodic table are
- Chemical symbol
- Atomic number (the number of protons)
- Atomic weight (the weight, in grams, of one mole of this element)

We'll talk more about moles in a later chapter. In chemistry class you are usually required to memorize the symbols for each element. Except for a few of them, most symbols are fairly easy to remember.

The next chapter is designed to help you make sure that you are proficient in working with the large numbers used in chemistry.

Chapter Two Review Questions

1. What is the definition of an atom?

_____

_____

_____

2. What is the definition of a molecule?

_____

_____

_____

3. What is your chair made out of, on a "micro" level?

_____

_____

_____

4. What were Leucippus' and Democritus' contributions to chemistry?

_____

_____

_____

5. How long would it take for you to count all the atoms in a tiny speck of aluminum foil (about 0.2 g—assuming you could see each atom)? *Hint: time yourself counting to 100.*

_____

_____

_____

6. Why is the Bohr model not correct?

_____

_____

_____

7. Why do chemists use the periodic table?

_____

_____

_____

Chapter Three

Scientific Notation

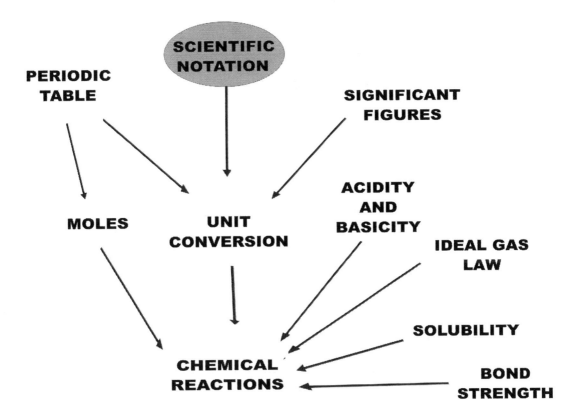

You may have already learned about scientific notation in school, in which case you might want to skim through this chapter (I suggest you test yourself with the review questions) and go on to the chapter on using your calculator.

Scientific notation is important for dealing with very large and very small numbers, as we often need to do in chemistry. Just like the periodic table, scientific notation is a tool to help you be more efficient. The reason we need to be able to work with huge or tiny numbers is the extremely tiny size of the atoms and molecules we are working with, as we saw in the chapter on the atom. You need to have an enormous number of atoms and/or molecules in order to get enough of them so that you can see and weigh them using a regular laboratory scale (hence the need to work with huge numbers).

When we are doing problems involving individual atoms or molecules, we also need to be able to use very tiny numbers because atoms, molecules and subatomic particles have such small weights. For example, the mass of an electron is equal to 0.000000000000000000000000000910938188 grams. That's pretty inconvenient to write down repeatedly, so it might be nice to have a simpler way to work with these numbers.

Scientific notation always consists of a number with a single digit to the left of the decimal point, and sometimes one or more digits following the decimal point, multiplied by 10 to some power. So the number 53 is shown as 5.3×10^1 and 0.005 is shown as 5×10^{-3}.

LARGE NUMBERS

Let's take the example from Chapter 1. We saw that the tiny piece of aluminum had about 45 quintillion or 45,000,000,000,000,000,000 atoms of aluminum in it, assuming it is pure aluminum. Scientific notation gives us an easier way to write this number:

45,000,000,000,000,000,000

can be shown as

4.5 x 10,000,000,000,000,000,000

If we isolate that second number,

10,000,000,000,000,000,000

can be shown as

10^{19}

Putting the two together:

45,000,000,000,000,000,000

can be shown as

4.5×10^{19}

Just count how many times you have to move the decimal point to the *left*, and stop when it is right after the first digit of the original number (in this case, on the right side of the 4). That number becomes the *positive* exponent attached to your "10." In this case, it was moved 19 times, giving us 10^{19}.

SMALL NUMBERS

For numbers less than one, such as 0.000000000501, count how many times you have to move the decimal point to the *right* to get to the right of the first non-zero number (in this case, on the right side of the 5) and put that number as a *negative* exponent on the "10." In this example, the decimal point was moved ten times.

0.000000000501

can be shown as

5.01 x 0.0000000001

0.0000000001

can be shown as

$$10^{-10}$$

So **0.000000000501**

can be shown as

$$5.01 \text{ x } 10^{-10}$$

If you try to get in the habit of writing most of your numbers in scientific notation as you work your chemistry problems, it helps you in the long run as you work problems.

The next chapter isn't really a chapter on a chemistry concept; instead, its purpose is to make sure you are using your calculator correctly to do problems using scientific notation. This is a key skill, so I highly recommend that you at least read it through and practice on a calculator.

Chapter Three Review Questions

1. Why is it so important to be able to use scientific notation in chemistry?

_____

_____

_____

_____

_____

_____

2. Can you just count the zeroes to get the right exponent for scientific notation? When does that work, and when does it give the wrong answer?

_____

_____

_____

_____

_____

_____

Chapter Four

Using Your Calculator

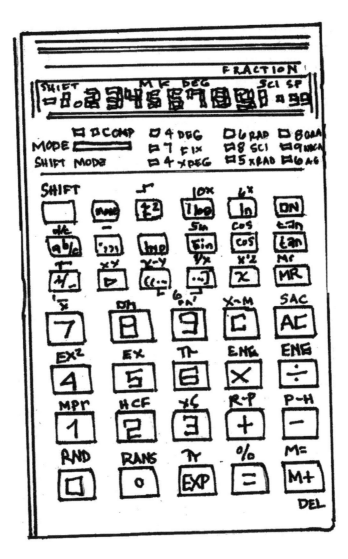

For chemistry, you are going to need a scientific calculator at the very least. This type of calculator does more than just add, subtract, multiply and divide. It also has many other functions like exponents and memory. Plus, there is a function just for doing scientific notation.

In order to use scientific notation, you have to either perform operations on exponents manually (doing it this way is simple but not very fast) or use your calculator. You should probably learn how to do it manually, because it comes in handy if you forget to bring your calculator to a test. However, I want to make sure you know how to use your calculator. I have seen students set up a chemistry problem correctly, then use their calculator incorrectly and get the wrong answer. This can definitely add to the frustration of learning chemistry!

I personally have never felt the need to buy one of the large, expensive graphing calculators—all that's really needed for chemistry is one that can perform the following functions (the keys to look for on the calculator are in parentheses):

- Scientific notation (**EE** or **EXP**)
- Logs (**log**) and natural logs (**ln**)
- Power and root functions ($\mathbf{x}^2$ and $\mathbf{x}^y$)
- Negative numbers (+/−)
- Trigonometric functions (**sin, cos and tan**)
- Inverses (**1/x**)

You may want to get a graphing calculator if you like them better and/or you think you might take calculus eventually, but I've been using the same ten-dollar scientific calculator for the past fifteen or twenty years, and it worked just fine for me all the way through graduate school, including many classes involving calculus. I would suggest asking your teacher for advice if you can't decide, but it is better if you start trying your calculator out over the summer to get used to it.

The following instructions should work for both simple scientific and graphing calculators. The instructions that come with a calculator are usually helpful, and there are plenty of tutorials on the Internet as well.

First, look on your calculator for a button that says either "**EXP**" or "**EE**"; this is the scientific notation button. In some cases you may have to press the "**2**nd" or "**Shift**" key first, if **EXP/EE** is not the primary function for that key.

VERY LARGE NUMBERS

To input **4.5 x 10^{19}**, for example, you enter the following:

Press **4.5** (four, decimal point, five).

Now hit the "**EXP**" or "**EE**" button.

One of the following should show up to the right of your 4.5:

"**00**" or "**E**" or "**10^**"

Enter "**19**"

It should look something like one of these three, depending on which company manufactured your calculator:

4.5   19

or

4.5   10^19

or

4.5   E19

On some calculators, the 19 is about half the size of the 4.5. These all mean the same thing; it's just that different calculator companies choose to show it in different ways.

Okay, now that you've input the number, you can add to it, multiply it, or perform any other operation. Try out a few operations just for practice—multiply this number by 2 or add 200 to it. You'll find that adding small numbers doesn't change it. You would have to add a number larger than 1.0 x 10^{20} to change a number this large so that you notice a change.

VERY SMALL NUMBERS

Now I'll show you how to do very small numbers. Let's use 6×10^{-20}.

On your calculator, you would enter this exactly as before, except this time you have to find a way to input –20 instead of 19. Use the same method for the first part:

Enter **6**

Press "**EE**" or "**EXP**"

Enter **20**

Now you look for a button that says "+/–". Press it.

A negative sign should come up before the number 20. The display looks like one of these:

6.   –20

or

6.   10^(–20)

or

6.   E–20

All of these are the same as writing 6×10^{-20}. Make sure you press the +/– button after pressing **EE** or **EXP**, so the negative sign is in the right place. Again, play around with this number. Notice that adding the number one to it, for example, just gives you one again because 6×10^{-20} is so small.

Now, if you practice these a few times on the calculator to get in the habit of using it for scientific notation, you will have an easier time in class and during exams.

Chapter Four Review Questions

1. What happens if you get confused and press the **10**^{x} key on your calculator instead of **EE/EXP**? How does this affect the answer to the problem you are trying to work?

_____

_____

_____

_____

_____

_____

_____

2. Can you solve scientific notation problems without using your calculator? How would you go about this?

_____

_____

_____

_____

_____

_____

_____

Chapter Five

Significant Figures

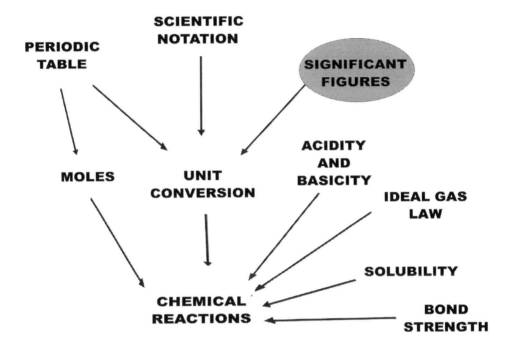

Chemistry is a relatively new science that grew out of the practices of al-chemists, whose main goal was to try to manufacture gold out of other metals, such as lead. Some of the alchemists eventually gave up on the search for gold and branched out into other materials, giving rise to a more mature field of scientific study.

In 1661, Robert Boyle wrote *The Sceptical Chymist*, which was the first written work drawing a clear line between alchemy and the science of chemistry. Boyle emphasized the necessity of designing experiments to test theories, instead of the haphazard and often mystical approaches favored by the alchemists.

Before the late 1700s, the experiments carried out by chemists involved mixing reactants together and noting the properties of the products. There was less of an emphasis on weighing products and reactants to measure *quantitatively* the substances produced. In other words, these chemists focused purely on the *qualitative* properties of matter, like color, appearance, odor and—yes—taste. Nowadays we know not to taste or smell laboratory chemicals, but until about a hundred years ago, it was routine to record the smell and taste of any new chemical produced, as well as its appearance and other physical properties. This led to the premature deaths of many of the early chemists.

Antoine Lavoisier (1743–1794), a French scientist, was the first person to emphasize quantitative measurement. He measured the weight of the individual substances that were mixed together before the chemical reaction had taken place, as well as the weight of the whole thing afterwards. No one had ever thought of doing this before. This led him to develop the law of conservation of matter, which states that matter is neither created nor destroyed, and that the mass of a closed system remains the same no matter what you do to it (as long as it remains closed).

For example, the weight of a reaction mixture decreases when gases are released during a reaction, unless you perform the reaction in a closed container. This may seem obvious to us today, but at the time knowledge about gases and atoms was very limited, so this kind of thinking was revolutionary. Unfortunately for the world of chemistry, Lavoisier was beheaded at age fifty during the French Revolution. Who knows what else he might have accomplished if he had had more time.

When we measure substances in chemistry, we like to know just how exact our measurements are. This is why we have something called significant figures. Take a simple measurement with a ruler. When you measure something, you look at the ruler's lines to see just how exact you can get.

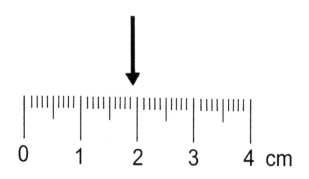

If you're looking at the metric ruler shown above, you can probably estimate down to 0.1 millimeter, since (if you squint) you can see the difference between something that measures 1.92 cm and something that measures 1.98 cm.

However, using this particular ruler, you can only really be *sure* about whether something measures 1.9 or 2.0 cm. So in this case, the first digit after the decimal point is the **significant figure** or **significant digit** (also sometimes called **sig fig** or **sig dig**), meaning that it is the one that has *importance*. In this case, the second number after the decimal point can't be trusted, which is why it isn't significant. If something measured 2 cm exactly using this particular ruler, we would say it measured 2.0 cm, but *not* 2.00 cm (and certainly not 2.000 cm!).

If you were trying to measure something and figure out whether it was 1.92 or 1.93, for example, you couldn't do it using the ruler shown above—you would need one with smaller divisions, maybe one you could only see under a microscope. You may have used one of those in biology class to measure cells or very tiny objects.

In chemistry, the usefulness of significant figures comes into play when you add, subtract, multiply or divide two or more measurements that were taken with different measuring devices, each of which has a different level of accuracy. The *least exact* measurement is the one that determines the level of accuracy for the final answer. The rules for addition/subtraction are different from those of multiplication/division. A few examples are shown below.

SIGNIFICANT FIGURES IN ADDITION AND SUBTRACTION

Let's say you want to know the combined weight of yourself and a friend. You weigh yourself with a metric bathroom scale (accurate to about the nearest 0.1 kg) and find that you weigh 62.2 kg. Then your friend weighs himself using a more accurate laboratory scale and finds that he weighs 61.256 kg. When you add the two measurements together, you can only keep the significant figures from the *less accurate* form of measurement, which in this case is the bathroom scale. Here is the procedure for either addition or subtraction:

STEP 1

Add the two measurements the way you normally would, lining up the decimal points:

62.2 kg
+ 61.256 kg
123.456 kg

STEP 2

Round to the nearest significant figure. For this example, the answer becomes 123.5 kg. The digit after the decimal point is the significant figure, because that was the *most* accurate digit for the *least* accurate measurement technique (the bathroom scale).

Sometimes you have to round an answer that has a final digit of 5. The rules for rounding these numbers vary. Some textbooks say to round up when the next digit to the left of the 5 is odd and round down if it is even. For example, 2.55 rounds up to 2.6, while 2.45 rounds down to 2.4. This method makes sense if you are collecting a large number of numerical data points during an experiment, because any rounding errors tend to even out in the end.

Other books tell you *always* to round up when 5 is the final digit, although this method is not as commonly used. My policy has always been to use the method the teacher recommends—it's hard to go wrong that way.

SIGNIFICANT FIGURES IN MULTIPLICATION AND DIVISION

For multiplying and dividing, the rule is to look at the measurement with the *lowest total* number of significant figures. Then you round the answer so that it has that same total number of digits. Always count the digits from the left to the right.

For example, say you take a measurement of the distance between your house and a friend's house. You find the distance to be 23.4 km (this measurement has 3 sig figs), but you want to convert this number into miles. You have a good conversion factor that has many significant digits (1 mile = 1.609344 km—this has 7 sig figs). In this case, you would use the total number of digits in the measurement *you* made for your answer, since that is the measurement that is the least accurate.

Here is the procedure for either multiplication or division:

STEP 1

Divide as you normally would to solve the problem.

23.4 km ÷ 1.609344 km/mile = 14.54009 miles (the km units cancel out)

STEP 2

Round it to 3 sig figs. Your answer becomes 14.5 miles.

In problems with multiple calculations, always complete all of the mathematical operations or conversions and only round the final answer.

Now let's go back to scientific notation for a minute. Scientific notation is useful because it allows you to show the number of sig figs even in large numbers with zeroes. For example, if you weigh out 1,800 kg of rocks, how do you show that you are only certain of this weight to the nearest 100 kg? You would use scientific notation to write it as 1.8×10^3 kg. Alternatively, if you were sure of the accuracy of the weight to the nearest 1 kg, you would write it as 1.800×10^3 kg.

Now that we have some of the basics down, let's move on to some actual chemistry concepts.

Chapter Five—Review Questions

1. What simple thing did Lavoisier do that revolutionized chemistry?

_____

_____

_____

2. When a gas is released in a chemical reaction, does the weight of the reaction mixture go up or down? Why?

_____

_____

_____

3. In what situation could you measure the weight of a chemical for a chemical reaction accurately, using an ordinary bathroom scale? Explain.

_____

_____

_____

4. Why is it important to know the accurate weights of reactants and products in a chemical reaction?

_____

_____

_____

5. When taking measurements using two devices, each with different accuracies, which one's significant figures do you use for your answer?

_____

_____

_____

Chapter Six

The Mole and Molarity

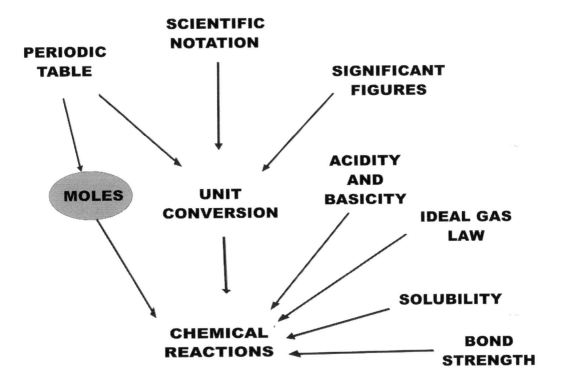

THE MOLE

The mole (often abbreviated as "mol") is a concept that I think some chemists make more difficult than it needs to be. With this in mind, I am going to try to start this chapter with something easy to remember and then explain it in more detail. So if you remember just one thing from this chapter, it should be this: *a mole is like a dozen*. In chemistry, one mole of atoms is equal to 6.022×10^{23} atoms. Just like a *dozen* bagels consists of 12 bagels, one *mole* of bagels consists of 6.022×10^{23} bagels.

The mole is a tool that chemists use for keeping track of the huge numbers of atoms and molecules that are necessary to make something big enough for you to see, like that piece of aluminum foil that we looked at in the atom chapter.

Why chemists use this number and not something that's easy to remember, like maybe 1.0×10^{50}, is a very good question. This number does actually have some historical significance, and it's hard to change a scientific constant once everyone starts using it. It's called Avogadro's number (symbol N_A), in honor of Amedeo Avogadro. In the early 1800s, he did some of the early work linking the volume of a gas with the number of molecules or atoms contained in that gas.

In the late 1800s, Jean Perrin, a physicist, proposed that the number of atoms or molecules in a mole should be set equal to the number of molecules of oxygen in 32 grams of oxygen gas (32 g O_2), which at the time was a standard amount that was called a "gram-molecule" of oxygen. One mole of oxygen atoms weighs 16 grams, so the weight of a mole of O_2 is double that weight. Most chemists agreed, the unit was shortened to "mole," and this constant has been used by chemists ever since. So basically we're stuck with Avagadro's number because of tradition. As with most traditions in the sciences, it's probably best to just accept it and move on with your life.

The periodic table shows you the grams per mole (g/mol) for every atom. These numbers have been determined by long-ago experiments and put into the periodic table for your use. You can add them up as you build a molecule to get the weight of one mole of any molecule you create. If you look at this book's periodic table in the earlier chapter on the atom, you can determine that 32 is the molecular weight of O_2 (16 grams/mole for each atom of oxygen x 2 atoms of oxygen).

Another interesting piece of information is that one mole of *any* gas occupies 22.4 liters at STP (standard temperature and pressure, defined as 0 degrees Celsius and 1 atm of pressure, approximately the atmospheric pressure at sea level).

When we talk about chemical reactions later in this book, the mole will be crucial, because it is the unit of proportion that you will use whenever you write out chemical equations.

MOLARITY

Another concept you will encounter in high school chemistry is molarity, which is moles per liter, usually shown as mol/L. This unit is used to express the concentration of solutions of *solutes*[3] dissolved in *solvents*.[4] The diagram below shows the important parts that make up a typical solution.

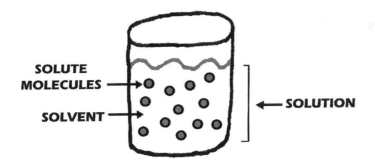

Molarity is another way of saying concentration, like when you are mixing up instant cake batter and you have to add the right amount of liquid to the cake mix for the cake to come out right. The molarity of the cake batter has

---

3   Solute–a substance dissolved in another substance, forming a solution. Sugar that is dissolved in water is an example of a solute.

4   Solvent–a substance, usually a liquid, capable of dissolving another substance. Water, alcohol, and benzene are examples.

to be correct, or you get a watery or lumpy mess instead of smooth batter that makes a good cake.

Molarity is used as a tool to figure out how much product you get when you combine a solution of one chemical with another chemical, because sometimes you can't buy the desired chemical in its dry form. We will cover an example of this concept in a later chapter on units and conversions.

Chapter Six—Review Questions

1. What is the one thing you should remember from this chapter?

_____

_____

_____

2. Why do chemists use moles?

_____

_____

_____

3. Where are molecular weight values found?

_____

_____

_____

4. What is the definition of a solution?

_____

_____

_____

5. What are the units for molarity?

_____

_____

_____

6. Why is it helpful to know the molarity of a solution?

_____

_____

_____

Chapter Seven

Units and Conversions

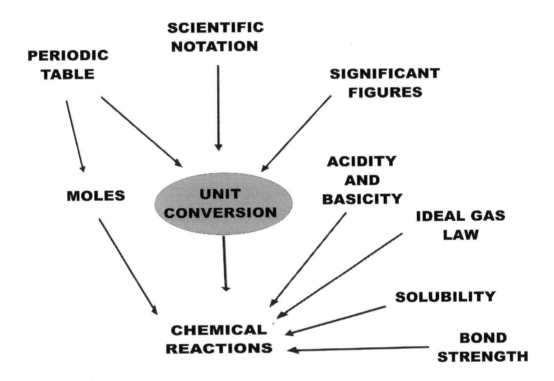

UNITS

Whenever we use numbers in chemistry, we are always talking about an amount of a substance, so there is always a *unit* attached to it. So if we're talking about one mole of a substance, we don't just write 6.022×10^{23}, we write

$$6.022 \times 10^{23} \textit{ bagels} = 1 \textit{ mole} \text{ of } \textit{bagels}$$

or

$$6.022 \times 10^{23} \textit{ Al atoms} = 1 \textit{ mole} \text{ of } \textit{Al atoms}$$

or

$$6.022 \times 10^{23} \textit{ H}_2\textit{O molecules} = 1 \textit{ mole} \text{ of } \textit{H}_2\textit{O molecules}$$

If you stick to writing down all the units, it makes your life a lot easier as you do your chemistry problems. It can seem like a pain in the neck at first, but it gets easier with practice. You should get in the habit of asking yourself "of what?" For example, "one mole of what?" or "one liter of what?" This helps you to keep track of what substance you are talking about as you perform each step of your conversions.

CONVERSIONS

In chemistry class, you will spend much of your time converting units. The methods in this chapter are the tools that will enable you to do simple unit conversions. They will also allow you to figure out how much product you will get when you perform a reaction, and how much energy is released. You have already done similar work in math class, when you converted grams to kilograms or miles to kilometers. It may have looked something like the equation below:

$$10 \text{ miles} \times \frac{1 \text{ km}}{0.62 \text{ mile}} = 16 \text{ km}$$

The process above is just a more sophisticated version of the "cross-multiplication" method you probably learned in elementary school. Notice that the "miles" units cancel out (like identical variables in algebra), leaving you with just kilometers. Also, notice that the numerator and the denominator of the conversion ratio are equal to each other. So in essence, you are just multiplying by the number one. Remember from fourth- or fifth-grade math that if the top and bottom are equal, then the fraction equals one:

$$\frac{3}{3} = 1$$

$$\frac{12 \text{ donuts}}{1 \text{ dozen donuts}} = 1$$

This is why the units that go with the numbers are very important. Otherwise, you're not multiplying by 1/1 and the conversion will be incorrect.

Whereas in your previous experience in math class you may have converted units like this:

Conversion

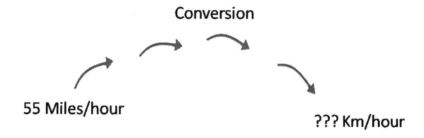

55 Miles/hour   ??? Km/hour

In chemistry we are going to need a more sophisticated tool for conversions like this one:

Simple Conversion

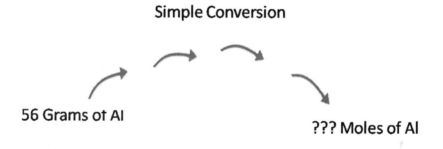

56 Grams ot Al   ??? Moles of Al

We need a system to help us keep track of all these units. In a moment, we'll see how that system works. Let's go back to that aluminum example. Assume that the little bit of Al foil weighs about 0.2 g.

Here's how you set it up. First, take the original measurement and put it on the left side of the page, then write down the units that you want for the answer on the right side of the page:

0.2 g Al   x   What Conversion Factors?   = ?? **atoms of Al**

Next, you try to figure out what you have to do to finish the conversion. It may take you a little time to find the right conversion factors so that you get the answer with the right units (in this case, atoms of Al). You just need to pick the right conversion ratios and flip them upside down if necessary to make the units cancel out. Remember that they are the same upside down or right side up, since the numerator and denominator are equal:

$$0.2 \text{gAl} \times \frac{1 \text{ mole of Al}}{27 \text{ g Al}} = 7.4 \times 10^{-3} \text{ moles of Al}$$

In this case, the conversion factor came from the periodic table—the number of grams in one mole of Al. Now we can go on to convert all the way to atoms, using Avogadro's number:

$$7.4 \times 10^{-3} \text{ moles of Al} \times \frac{6.022 \times 10^{23} \text{ atoms of Al}}{1 \text{ mole of Al}} = 4.5 \times 10^{21} \text{ atoms of Al}$$

Again, just think of these conversions as multiplying by 1/1, and that should help you remember that the amounts on the top and bottom of the ratio you use to convert always have to be equal to one another. We can actually put these two conversions together and cancel out the units to get atoms of Al:

$$0.2 \text{ gAl} \times \frac{1 \text{ mole of Al}}{27 \text{ g Al}} \times \frac{6.022 \times 10^{23} \text{ atoms of Al}}{1 \text{ mole of Al}} = 4.5 \times 10^{21} \text{ atoms of Al}$$

Notice how the units cancel out neatly (because of the way you set it up), leaving you with the units "atoms of Al." That's how I figured out how many atoms were in that little bit of aluminum foil back in the chapter on the atom.

Sometimes you need to convert a few units at once, like kg/L to mg/mL. The conversion method allows you to do this, too. Let's take an example of a conversion from units of kg/L to mg/mL:

$$0.1 \frac{\text{kg NaCl}}{\text{L H}_2\text{O}} \times \text{What Conversion Factors?} = \text{?? mg NaCl/mL H}_2\text{O}$$

You want to get from kilograms of NaCl per liter of H_2O to milligrams of NaCl per milliliter of H_2O. This is a little more complicated, but it's not too scary if we break it down into pieces. Let's look at each unit separately and fill in the blank with conversion factors:

- Since "L H_2O" is in the denominator of the original number, and the units in the denominator of the answer are "mL H_2O," we want to look up a conversion factor for L to mL and put the "L H_2O" units in the numerator and "mL H_2O" in the denominator
- Use this conversion factor: 1 L H_2O = 10^3 mL H_2O
- Do the same for the second conversion of kg NaCl to mg NaCl. You want the "kg NaCl" to be in the denominator and mg NaCl in the numerator so everything will cancel out correctly
- Use 1 kg NaCl = 10^6 mg NaCl

$$0.1 \; \frac{kg \; NaCl}{L \; H_2O} \; \times \; \frac{1 \; L \; H_2O}{10^3 \, mL \; H_2O} \; \times \; \frac{10^6 \; mg \; NaCl}{1 \; kg \; NaCl} \; = \; ?? \; mg \; NaCl/mL \; H_2O$$

Now we cancel out the matching units to get the final answer:

$$0.1 \; \frac{\cancel{kg \; NaCl}}{\cancel{L \; H_2O}} \; \times \; \frac{1 \; \cancel{L \; H_2O}}{10^3 \, mL \; H_2O} \; \times \; \frac{10^6 \; mg \; NaCl}{1 \; \cancel{kg \; NaCl}} \; = \; 100 \; mg \; NaCl/mL \; H_2O$$

It makes the problems easier if you actually cross out the units with your pencil as you go, as shown above.

Alternatively, we might have to figure out how much product is created by a reaction. This is similar to other conversions, with the added complication of using a chemical reaction to convert from moles of reactants to moles of product:

Conversion + Reaction with Cl

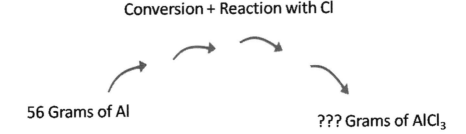

56 Grams of Al   ??? Grams of $AlCl_3$

In a later chapter, you will see how the use of this same conversion method enables you to figure out how much product you will get from a chemical reaction.

Also, sometimes in chemistry we say cm^3 (or cubic centimeters) instead of mL. They both mean the same thing—for example, 40 cm^3 or mL is the "40 cc" you hear them talking about on medical shows or at the doctor's office when someone's about to get an injection. When converting from cm^3 to m^3, it is easier to set up the conversion like this:

$$40 \text{ moles NaCl/m}^3 \text{ of } H_2O \quad \times \quad \frac{(1 \text{ m})^3 \, H_2O}{(100 \text{cm})^3 \, H_2O} = 4 \times 10^{-5} \text{ moles NaCl/cm}^3 \, H_2O$$

Now if you haven't seen this kind of problem before, you might be asking yourself how I got 10^{-5}, since 40 divided by 100 is 0.4. This is a common mistake in chemistry class—cubing the units and forgetting to cube the number. The entire value inside the parentheses is cubed, so you are dividing by "$10^6 \, cm^3$," *not* "100 cm^{3}". Just something to be careful about as you work your conversion problems. Here's another tip:

If you can make your units cancel so the answer has the correct units, you have a better chance of getting the right answer in a chemistry conversion problem.

This is actually as simple as it sounds. If you're stuck on a test, try something that gets the units to cancel so that your answer has the right units, and you've increased your chances of getting it right. It doesn't magically give you the answer, but it might at least get you partial credit or get you on the right track towards finding the right answer.

Chapter Seven—Review Questions

1. Why is it important to attach units to numbers when working chemistry problems?

_____

_____

_____

_____

2. What must be true about the numerator and the denominator of a conversion factor?

_____

_____

_____

_____

3. How does setting up your conversions so that the units cancel help you to get the correct answer to a chemistry problem?

_____

_____

_____

_____

Chapter Eight

Acidity and Basicity

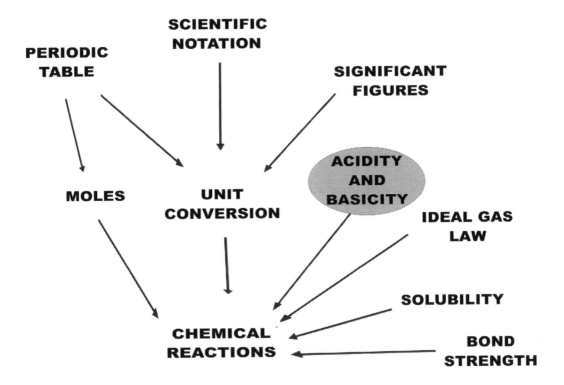

cids and bases are compounds that have a special kind of reactivity. You have already encountered acids and bases in your everyday life. Your stomach uses acid to dissolve the proteins you eat. You know that lemon juice contains acid, because it tastes sour. Sodium bicarbonate is a weak base that you may have seen used to make a "volcano" for a science project in grade school by reacting it with vinegar (acetic acid).

First, let's talk about the H^+ ion, also called a proton. Why do we call it a proton? If you look back at the periodic table in the Atoms chapter, you'll see that while helium (He) has two protons and two neutrons (giving it an atomic weight of four), hydrogen only has an atomic weight of one. What's going on here? Why isn't its weight equal to two (one proton and one neutron)? Hydrogen is a special element because the most abundant form found on earth only has one proton and one electron, but no neutrons.[5] This means that when you strip away that one electron to get H^+, what you have left is actually just a proton. In chemistry class, therefore, *chemists use the names "H^+" and "proton" interchangeably.*

The Brønsted-Lowry theory of acids and bases defines an acid as a molecule that is able to *donate one or more protons (H^+ ions)*. Bases are defined as molecules that are able to *accept protons*. This theory was developed in 1923.

The other acid-base theory is called the Lewis theory[6], which defines the acid as a molecule that can *accept an electron pair*, and a base as a molecule that can *donate an electron pair*. This definition is more generally useful, since it covers not only molecules that have protons, but also molecules that are acids but do not have any protons. In this book, however, we will focus on Brønsted-Lowry acids and bases because they are easier to understand when you're just starting out.

Brønsted-Lowry acids and bases work together to move protons from one place to another; every acid has a base associated with it (called its *conjugate base*) and every base has a *conjugate acid*. Here are two examples:

- The conjugate base of HNO_3 (nitric acid) is NO_3^- (nitrate ion). In water, HNO_3 comes apart to form H^+ and NO_3^-.
- The conjugate acid of the base ammonia (NH_3) is NH_4^+ (ammonium ion). In aqueous (water) solution, the base NH_3 reacts with H_2O, removing a proton from the water to form NH_4^+ and OH^- ions.

The pH scale is a way to show acidity and basicity. You probably already learned about the pH scale in your previous science classes. Remember that acids have a pH less than 7, while bases have a pH greater than 7. Water, which is formed when H^+ and OH^- are combined, has a perfectly neutral pH of 7. We use this number because it has been found experimentally that pure water is

---

5   There are also two other types (called *isotopes*) of hydrogen with 1 and 2 neutrons (called deuterium and tritium, respectively), but they are not nearly as abundant on earth, and so I will ignore them for the purposes of this book.

6   This theory was also first reported, oddly enough, in 1923.

naturally slightly ionized, containing about 10^{-7} moles/L of H^+ ions and 10^{-7} moles/L of OH^- ions. Søren Peder Lauritz Sørensen defined the pH scale in 1909 as the negative logarithm (base 10) of the concentration of H^+ ions:

$$-\log (10^{-7}) = 7 = pH \text{ of pure water}$$

Scientists are divided over what Sørensen meant the "p" in pH to stand for—some say pH was meant to stand for "power of Hydrogen." It doesn't really matter what it stands for, but the word "power" may help you to remember that exponents are involved.

The pH scale is *logarithmic,* which means that acidity increases dramatically as the pH goes down. Therefore, a pH of 3 is ten times more acidic than a pH of 4, pH 2 is ten times more acidic than pH 3, and so on. The Richter scale, used to measure the severity of earthquakes, has a similarly exponential scale. An earthquake of 7.0 is ten times stronger than an earthquake that measures 6.0, a 6.0 is ten times stronger than a 5.0, and so on down the scale.

If you add enough acid to some amount of pure water so that the concentration of H^+ ions in the solution is now ten times greater—then the pH goes *down* by one, because $10^{-7} \times 10 = 10^{-6}$, which is greater than 10^{-7}. The pH would change from 7 to 6 because $-\log (10^{-6}) = 6$.

If the concentration of H^+ ions is *decreased* in an acidic solution by adding a base to react with the H^+ ions (since $H^+ + OH^-$ forms H_2O), the pH goes *up.* Again, you would have to increase the concentration of OH^- ions by a factor of ten in order to increase the pH by one.

Controlling acidity and basicity is very important in chemical reactions. Many of the reactions that go on in your body are acid/base reactions. If problems with these reactions develop, it could cause you some serious trouble. You may have heard of someone having an electrolyte imbalance; this is the medical term for acid/base problems in the body. This condition is usually caused by dehydration and/or poor nutrition.

You will balance many acid/base reactions in chemistry class, so it is helpful for you to know the practical reality of these substances and their reactions. In the chapter on chemical reactions (the final chapter in this book), I give an example of an acid/base reaction.

Chapter Eight—Review Questions

1. What are two acidic substances that you can think of that you have encountered in your everyday life?

_____

_____

_____

2. Can you name any commonly used bases? What are they used for?

_____

_____

_____

3. Do bases have a pH less than or greater than 7?

_____

_____

_____

4. What does "pH" stand for?

_____

_____

_____

5. Are there any acids or bases in your body? What are they?

_____

_____

_____

6. What is an electrolyte imbalance?

_____

_____

_____

Chapter Nine

The Ideal Gas Law

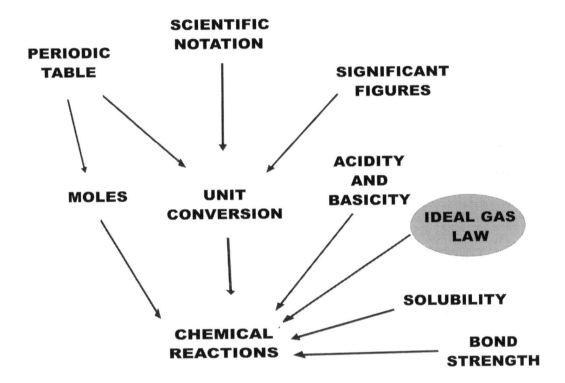

An ideal gas is a theoretical gas whose molecules bounce off each other or the walls of their container, but never gain any energy from these collisions. Each molecule or atom's size is also considered to be so small as to be unimportant.

At room temperature, many gases behave pretty much like an ideal gas, or at least close enough for our purposes. The ideal gas law is a key concept in chemistry class. It is used to determine how many moles of a gas are contained in a container of a certain volume, at a certain temperature (this is referred to as a *system*). This law is useful for determining how much gas to add to a chemical reaction, if one of the reactants is in the form of a gas. It can also be used to determine how much gas is produced by a reaction. In addition, if you know the number of moles of the gas, this equation can allow you to determine the other physical properties of the molecules in the system (like temperature, volume, or pressure).

The equation associated with this gas law is **PV = nRT**, where

- P is the pressure the gas exerts on its container
- V is the volume occupied by the gas
- n is the number of moles of the gas
- R is the gas constant (8.314 J/(mol K))
- T is the temperature in degrees Kelvin (K)

At first glance, plugging numbers into PV = nRT may seem easy, but don't let it fool you. Unless you study the concepts behind this equation, you will have a lot more trouble with the harder problems that come later.

SOLVING IDEAL GAS PROBLEMS

In chemistry class, you will start out with problems that give you all of the variables in the ideal gas law equation except one; then you have to solve for that one variable. More advanced problems involve two-variable equations—like the kind you solved in algebra class using x and y. Knowing the basic ideal gas law equation gets you started, but if you start having problems, make sure you go back and practice some algebra using this equation.

For example, you may encounter problems that deal with the beginning and ending state of a system. You may see something like the following, where the subscript 1 refers to the beginning state of the system, and the subscript 2 refers to the end state (after you have done something to it, like heat it up or apply more pressure):

$$\text{If } P_1V_1 = nRT$$

$$\text{and } P_2V_2 = nRT$$

$$\text{Then } P_1V_1 = P_2V_2$$

If you have trouble with this kind of logical math equation, I suggest that you definitely review some algebra with a chemistry or math teacher.

As you read the examples that follow, keep looking at the equation PV=nRT and try to see how the math matches up with the description.

PRESSURE AND VOLUME (*PV* = nRT)

When you squeeze a balloon, what happens? You feel it push back against your hand. At a molecular level, the pressure of a gas is caused by all those molecules of nitrogen and oxygen hitting the walls of the balloon and bouncing back again. There are so many molecules and they are so small that you cannot feel the individual collisions; you just feel steady pressure.

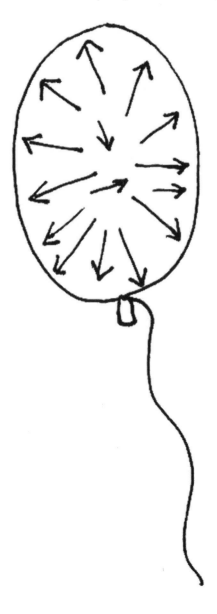

As you squeeze the balloon, you are lowering the volume of the balloon, which increases the pressure of the gas inside. Another part of the balloon also stretches out to release some of the pressure.

Imagine that all the molecules in the balloon are people in your hallways at school. In between classes, when the halls are crowded, you're much more likely to bump into someone else as you walk along than if there are only one or two people walking (increased number of moles of people causes more collisions, which causes increased pressure).

Now imagine that the hallway walls start closing in on you (decreased volume causes more collisions)—you would bump into other people more and more, and probably get shoved into the wall a few times too, as you tried to get to your next class. This is what's happening to the molecules in that balloon when you squeeze it and decrease the volume. They are hitting each other with more energy and banging into the walls, which you feel as pressure against your hand as you squeeze the balloon.

P and V are *inversely proportional*—as one goes down, the other goes up. If you *decrease* the volume of a container with a moveable wall, the pressure of the gas inside *increases*. And if you *increase* the volume of the same container, the pressure of the gas inside *decreases*.

Think of a closed syringe (without a needle attached). If you put your finger over the hole on the end and put pressure on the plunger, the volume will decrease, and you will feel the pressure against the plunger increase—volume down, pressure up. If you pull on the plunger, you will feel the lower pressure inside the syringe pulling against the plunger—volume up, pressure down.

Looking at the equation PV=nRT, if we keep temperature (T) and number of moles (n) constant during a change in the system, then:

- $P_1V_1 = nRT = P_2V_2$, so $P_1V_1 = P_2V_2$ when T and n are constant
- If the initial volume (V_1) *decreases* as the system goes to its final volume V_2 (when you press in the plunger), then the pressure P_1 has to *increase* as it goes to its final pressure P_2, in order for the left and right sides of the equation to remain equal

TEMPERATURE (PV = nR*T*)

Temperature is ***directly proportional*** to both the pressure and the volume of a gas. So when you increase temperature, either the volume or the pressure have to increase too. As temperature goes up, the atoms/molecules start moving around faster and faster. Again, think about those hallways in your school. If everyone starts running instead of walking to class, the number of collisions between you and other people definitely increase.

Another example is the tires on a car. In the summer, the tire pressure increases, while in the winter it decreases. Therefore, the tire becomes slightly more inflated under hot conditions. Increased temperature leads to increased pressure on the tires, which causes the volume of the gas to increase slightly to try to relieve the pressure. If a car catches on fire, the tires explode

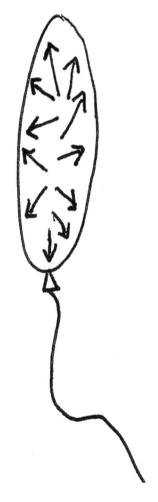

because the intense heat increases the pressure inside until the tire walls rupture because they can no longer hold the enormous pressure of the heated gases.

The opposite is also true—as the temperature goes down, both the pressure and the volume of the gas decrease. If you put an inflated balloon in the freezer for about 15 minutes, it looks partially deflated when you take it back out. The air stayed in the balloon; the atoms just slowed down from the cold, so they're not bumping into the balloon walls as fast or as often. This reduces the pressure on the balloon walls, causing deflation.

When you take the balloon out of the freezer, after about 15 minutes it warms back up and the balloon goes back to its original size. Again, increasing the temperature makes the pressure increase, which then causes the volume of the flexible balloon to increase until the pressure becomes the same as the surrounding atmosphere.

NUMBER OF MOLES (PV = *n*RT)

The number of moles of gas is (like temperature) directly proportional to the pressure and the volume. As the number of moles is increased, either the pressure or the volume (or both) increase. Think about blowing up a balloon, and imagine the gas molecules rushing into the balloon as you inflate it. Because the balloon is flexible, first the pressure and then the volume of the balloon rise, until the pressure inside the balloon is close enough to the surrounding atmospheric pressure that its pressure is no longer able to move the balloon's walls out further.

If we keep that volume constant (for example, if we put the balloon in a box before blowing it up), the pressure of the trapped gas rises as the number of moles of gas increases. If we could keep the pressure constant by having a super-flexible balloon, then only the volume would go up when the number of moles went up—the pressure would never increase.

In the next chapter, we will go back to solids and solutions and learn about another concept that is useful when performing chemical reactions.

Chapter Nine—Review Questions

1. What is the main equation associated with the ideal gas law?

2. What causes pressure on the walls of a container, at the molecular (micro) level?

3. Why is it important to be proficient in algebra so that you can solve ideal gas problems?

Chapter Ten

Solubility

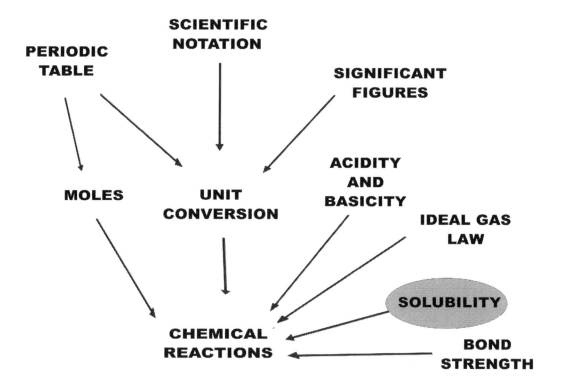

Think for a moment about the dry cleaning process for clothing. How does it work? Does the dry cleaning process really use something dry, like powder? Well, dry cleaning is dry in the sense that it is water-free, but it isn't liquid-free.

What happens when you wear your clothes is that the oils from your skin build up on the clothes, trapping dirt in the clothing. Bacteria start multiplying and digesting the oils, giving off foul-smelling waste products, and eventually you need to wash them. Oils are **hydro*phobic***, which means "water-fearing." Plain water will not remove these oils from your clothes. On the other hand, molecules that are soluble in water are said to be **hydro**ph$ilic$, which means "water-loving."

You may have tried pouring oil into water and found that the two liquids do not mix. The oil might go beneath the surface of the water for a few seconds because of gravity, but it quickly settles into a layer above the water. Water is denser than oil, which is why the oil stays on top.

If you put a few drops of oil into a glass of water, you will see that the oil stays together in globs at the surface. This is not so much because the oil molecules are attracted to one another, as it is that they are "afraid" of the water surrounding them, or hydrophobic. They push towards each other to get as far away from the water as possible, and end up crowding together with each other on top of the water.

Since clothes labeled "dry clean only" are ruined by immersion in water, you can't just throw them in the washing machine. You have to take them to the dry cleaner's. The dry cleaning process uses a hydrophobic solvent, usually some kind of chlorinated carbon-based chemical like trichloroethylene (also called perchloroethylene). The clothes are agitated in this solvent, until the solvent has dissolved all the oils left by your body and rinsed them away. The clothes never get wet with water, so they don't shrink or twist out of shape. When the dry cleaning solvent becomes too dirty, it is recycled (separating the pure solvent from the dirt and oil by distillation) and then reused.

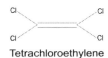

Tetrachloroethylene

With normal clothes, you would put them in the washing machine with soap, which has both hydrophobic and hydrophilic properties. The soap dissolves the oil and makes it water-soluble so that it can be rinsed away by the water.

A soap molecule is shown below, drawn in chemistry shorthand. The hydrophobic end of the molecule can dissolve oils, while the hydrophilic end dissolves in the surrounding water. When the clothing is rinsed, the water dissolves the hydrophilic end, which then of course drags along the hydro-phobic end of the soap that has the oils and dirt attached.

This end is water-soluble   This end is oil-soluble

There are lots of different types of detergents and soaps that basically function in just the same way as this molecule. Laundry detergent companies spend a lot of time researching better ways to dissolve the biggest variety of dirt, stains, and oil out of your clothing.

A good phrase to remember for solubility is *like dissolves like*. Hydrophobic liquids (such as dry cleaning solvent) dissolve other hydrophobic liquids (such as skin oils), and hydrophilic liquids (such as water) dissolve other hydrophilic substances (such as table salt).

Chapter Ten—Review Questions

1. What do hydrophobic and hydrophilic mean in chemistry?

_____

_____

_____

2. How does dry cleaning get clothing clean?

_____

_____

_____

_____

_____

3. How do soap and water get clothing clean?

_____

_____

_____

_____

Chapter Eleven

Bonding

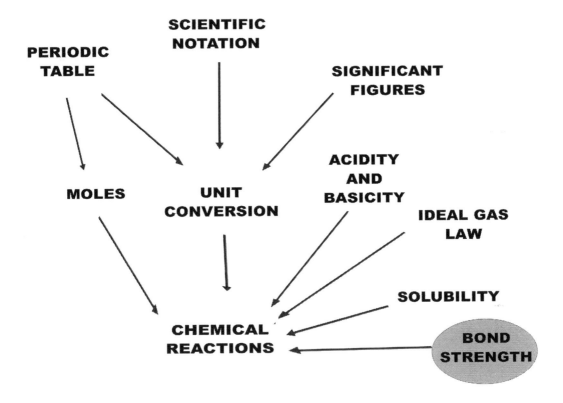

This chapter covers some basic concepts involving bonding and the release of energy associated with breaking chemical bonds. Bonding is how atoms "stick together" to form molecules, and it is one of the main reasons chemistry is so useful; knowing the strength of chemical bonds tells you how much energy is stored within a molecule. Knowing what types of bonds are in a molecule also gives clues to what the molecule can do or what other molecules can react with it.

Right now, your body is breaking and forming countless numbers of chemical bonds in order to generate the energy you need to be able to move around and stay alive. Studying these bonds is one way scientists figure out how the body works and how to cure diseases.

THE OCTET RULE

Atoms tend to follow the octet rule. This means that they want eight electrons (an octet) in their outer shell of electrons, which is known as the valence shell. In the valence shell, the electrons are the farthest from the nucleus of the atom and are most likely to interact with the electrons of other atoms. Some smaller atoms, like hydrogen, only want two electrons in their outer shell, but that is because they have so few protons, so they don't have enough of a positive charge to hold onto eight negatively-charged electrons. To attract larger numbers of electrons, the positive charge on the atom's nucleus has to be big enough. An atom with three protons, for example, could never hold onto eight electrons (–8 charge) with its puny +3 charge.

When drawing molecules and figuring out how many electrons an atom needs to make it happy (atoms are "happy" when they are at their lowest energy state, i.e., stable) it helps to draw what are called electron dot structures. These are very useful for keeping track of your electrons. For example, the structure for water looks like this:

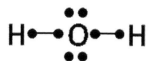

Let's just look at the oxygen (O) atom first. Notice that the electrons closest to the oxygen atom add up to six. This means the oxygen atom is neutral, with its positive and negative charges exactly balanced out. How do I know this? I looked at the periodic table and found that oxygen is in the sixth column, counting from the left hand side. This means that it has six valence electrons in its outer shell when it does not have a charge on it.

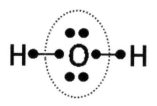

If you count the valence electrons plus the electrons on the other end of each bond with hydrogen, it adds up to eight.

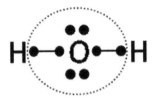

Eight is the "magic number", so you know the water molecule is happy, although not necessarily neutral. If there were only five valence electrons on the oxygen atom, you would have to assign the molecule a positive charge. If there was one extra electron, there would be a negative charge. Atoms become charged when the number of protons and valence electrons are un-equal. I suggest you try drawing a few of these structures for atoms and then for some of the molecules that have been discussed so far in this book (NH_3, NH_4^+). Notice which molecules end up with lone pairs (like the non-bonding electrons shown above and below the oxygen (O) atom in the water diagram above).

Most of the molecules you will encounter in high school and introductory college chemistry follow the octet rule. But of course, every rule has an exception. There are situations where the octet rule does not apply, but that is not covered in this book because it involves more advanced chemistry. Just be aware that the octet can sometimes be expanded to ten or more electrons for some of the larger atoms.

If you continue with your chemistry education, the electron dot structure method (also known as the Valence Shell Electron Pair Repulsion (VSEPR) model) will help you figure out many types of reactions, especially in organic chemistry (the study of carbon-containing molecules).

BONDS

In chemistry, we usually show the amount of energy needed to form a bond, or the energy given off when a bond is broken, in units of Joules (J) or Calories (Cal). Since in the real world we are usually talking about the amount of energy given off when *many* bonds are broken in a large number of reacting molecules, we often use units of J/mole or Cal/mole to show energy changes.

A strong bond is called a high-energy bond, and a weak bond a low-energy bond. There are many different kinds of bonds, and the differences between the types of bonds boil down to what the electrons are doing. Bond strength also determines whether a substance is a solid, a liquid, or a gas at room temperature. Here is a scale showing approximate relative bond strengths:

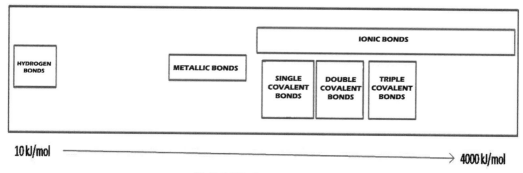

BOND STRENGTH

What you should take from this picture is just a general idea of the relative strength of the different bond types. Notice that ionic bonds can be as strong—as or stronger than—covalent bonds. There is no way to determine how strong a bond is just by looking at what type of bond it is—the strength of each bond is experimentally determined by performing various reactions and measuring the energy that is used up or given off by the reaction.

What is the difference between a covalent and an ionic or metallic bond? And what about those other bonds on the scale? Let's look at each one in a little more detail.

IONIC BOND

In an ionic bond, one or more electrons are actually "stolen" by another atom. This is similar to symbiotic relationships, which you learned in biology class. One atom wants an extra electron; the other atom has one that it wants to lose, so they both benefit.

Atoms that participate in ionic bonds generally want to get the same number of outer-shell electrons as one of the noble gases (helium (He), neon (Ne), etc.), which are shown on the far right side of the periodic table.

Electrons around an atom have energy levels, which are sort of like an onion's layers. Noble gases such as Ne have eight electrons in their outer shell to match their eight protons, which makes them extremely low in energy and very unlikely to react with any other atoms. Any other atoms besides the noble gases have to lose or gain an electron (i.e., pick up a positive or negative charge) to get into that eight valence electron state. Here's the exception— He only has two electrons total, so Li and Be will only try to have a total of two electrons in their outer shell (forming Li^+ and Be^{+2}).

The schematic below shows how an ionic bond is formed.

Na   Cl

Both sodium and chlorine are neutral, but would like to
have the same number of electrons as the nearest noble gas

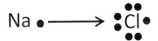

Sodium loses an electron to chlorine to get to the electron
configuration of Neon. Chlorine gains an electron to get to
the electron configuration of Argon

Na^+ $:\overset{\bullet\bullet}{\underset{\bullet\bullet}{Cl}}:^-$

Na now has one   Chlorine now has
less electron, and   an extra electron,
has a positive   so it has a
charge   negative charge

Na^+ Cl^-

The two are attracted to each other because of their
opposite charges, forming an ionic bond.

Sodium (Na) is on the far left of the periodic table. If it can get rid of one electron, it will have the same number of electrons in its outer shell as Ne. Chlorine (Cl) is in the next to last column, on the right side of the periodic table. If it can pick up one extra electron, it can get the same number of outer-shell electrons as argon (Ar).

Once the atom has transferred, these two atoms (now ions) have opposite charges (Na$^+$ and Cl$^-$), so they are attracted strongly to each other. The connection they form is called an ionic bond.

Incidentally, there is a balance between an atom's need to have eight electrons to complete its outer shell and its need to be neutral. Losing or gaining one or two electrons is usually fine, as long as the atom still gets to have that octet. Losing or gaining more than three electrons makes the atom very unstable, so this is very unlikely to happen for most atoms, even if it means it can complete its octet.

COVALENT BOND

A covalent bond is one in which the electrons are *shared* between two atoms. Instead of the electrons just staying close to one atom's nucleus, they can be associated with more than one nucleus. If we could freeze the electrons and take a picture, a bond between hydrogen and fluorine might look something like this:

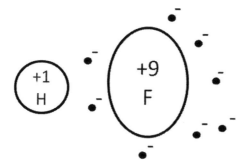

Hydrogen (H) is the smaller nucleus and fluorine (F) is the larger nucleus. Two of the eight total electrons are shown being shared between the two nuclei. In standard chemistry notation, this would be shown as H–F. Stronger versions of this bond are called double and triple bonds and involve more complicated sharing arrangements that store more energy.

Hydrogen cyanide, HCN, is a strong poison because of its high-energy triple bond, which releases a large amount of energy in the body if it is ingested. This bond would look more like the figure below. Nitrogen has two electrons that are not participating in this molecule's bonding. In chemistry shorthand, this molecule is shown as H–CN.

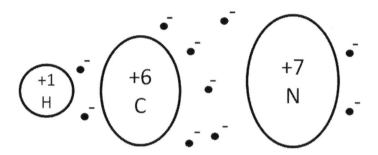

HYDROGEN BOND

A hydrogen bond is a kind of dipole-dipole[7] bond between the hydrogen and another atom that has a partial negative charge on two different molecules. The best example is in water. The hydrogen in a water molecule has a slightly positive charge and oxygen has a slightly negative charge, so they are very weakly attracted to each other. It looks something like this:

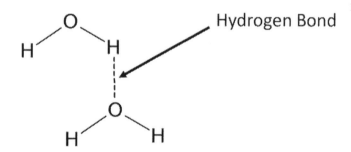

METALLIC BOND

A metallic bond is what holds together positively charged ions of metal atoms (positively charged ions are called *cations*, and negatively charged ions are called *anions*). Pure metals exist as a collection of metal cations, surrounded

---

7 Dipole—a situation where the electrons shared between two atoms move more toward one of the atoms, causing a slightly positive charge on the atom with fewer electrons and a slightly negative charge on the atom that ends up with more electrons.

by a "sea" of electrons. So really, when we talk about a metallic bond, we are talking about the average of a large group of bonds. If we could see it, it might look something like this:

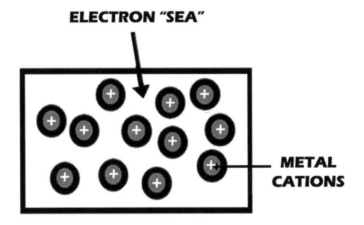

Bonds are important in chemistry because they provide the energy need-ed to drive a reaction forward towards the products. Thinking of bonds as groups of electrons that are always in motion should help you to visualize and understand the behavior of atoms and molecules more effectively.

Chapter Eleven—Review Questions

1. What is the octet rule, and why is it useful?

2. Which types of bonds involve shared electrons?

3. Which types of bonds involve "stolen" electrons?

4. Explain what a dipole-dipole bond is, and which type of bond covered in this chapter is a dipole-dipole bond.

Chapter Twelve

Chemical Reactions

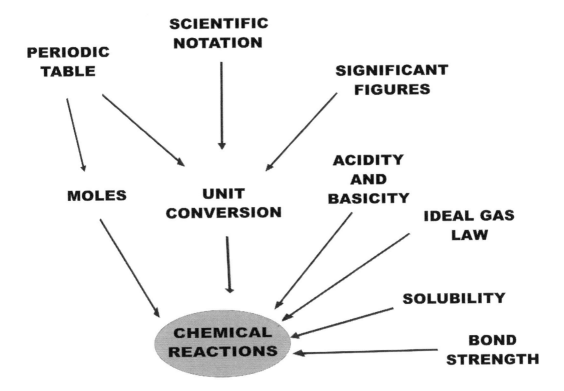

Now we come to the fun part—chemical reactions. This is where chemistry gets interesting and where you get to put some of those tools you've learned to good use. You'll get into reactions a lot more in class, but in this book I just want to show you how reactions fit in with the other concepts we've been talking about.

Here is an example of a problem that can be solved by using knowledge of chemistry (although this problem would probably not come up outside a chemistry lab). Let's say you have a liter of old hydrochloric acid (HCl) and you want to neutralize it so it won't hurt anyone. How do you know how much acid neutralizer to use? Remember the mole? It comes in handy when we want to do a reaction like this one, because when you're reacting one chemical with another, you want to know how many molecules of each reactant you need to get a good product (or any product at all!). This is because reactions happen at the micro level, but as I said earlier it is easier to measure them at the macro level, where the amounts of the substances involved are visible to the naked eye.

In essence, we're trying to take each and every H^+ ion in the reaction mixture and react it with an OH^- ion. This adds up to many, many individual reactions. For example, suppose we have one mole of HCl molecules and we want to neutralize it using a base, like sodium hydroxide (NaOH). We are going to need one mole of OH^- ions (from the NaOH) to react with the mole of H^+ ions (given off by the HCl). Both NaOH and HCl break apart into ions when dissolved in water.

We need the same number of NaOH molecules as HCl molecules to match the molecules up one-to-one. The reaction products are water and a salt (H_2O and NaCl). Remember, NaOH and HCl have different molecular weights, so this isn't a question of just adding an equal weight of NaOH to the HCl. You need to use moles to account for the difference in molecular weight between the two molecules.

Here's how we show this kind of reaction. Each number represents the number of moles of that molecule. Generally, it is not necessary to show the "1" for 1 mole, but I've included the numbers for clarity:

$$1 \text{ HCl} + 1 \text{ NaOH} \longrightarrow 1 \text{ H}_2\text{O} + 1 \text{ NaCl}$$

Incidentally, the process of figuring out the number of moles you need for a reaction to go to completion is called *stoichiometry*.

The equation above is just an easier way of writing this statement:

6.022×10^{23} molecules of HCl + 6.022×10^{23} molecules of NaOH

6.022×10^{23} molecules of H_2O + 6.022×10^{23} molecules of NaCl

The equation shown above was a very simple one—things can certainly get more complicated. For example, suppose you decided to use calcium hydroxide, whose formula is $Ca(OH)_2$, instead of sodium hydroxide. You only need half as many moles of $Ca(OH)_2$ to neutralize the HCl, since each $Ca(OH)_2$ molecule releases two OH^- ions as the reaction proceeds. Another way to say it is that half of one mole of $Ca(OH)_2$ contains 6.022×10^{23} OH^- ions:

$$1\ HCl\ +\ 0.5\ Ca(OH)_2\ \longrightarrow\ 1\ H_2O\ +\ 0.5\ CaCl_2$$

Again, this is just shorthand. The following is the complete version:

$$6.022 \times 10^{23}\ \text{molecules of HCl} + 3.011 \times 10^{23}\ \text{molecules of } Ca(OH)_2$$

$$6.022 \times 10^{23}\ \text{molecules of } H_2O + 3.011 \times 10^{23}\ \text{molecules of } CaCl_2$$

The useful part comes when you're actually trying to figure out how many grams of the base to weigh out to do your reaction. Let's use $Ca(OH)_2$ as the base for this example. You can use your conversion tools for this one—if you need 0.5 moles of $Ca(OH)_2$, then:

$$0.5\ \text{mole } Ca(OH)_2 \times \frac{74\ \text{g } Ca(OH)_2}{1\ \text{mole } Ca(OH)_2} = 37\ \text{g } Ca(OH)_2$$

So now you can weigh out 37g of the $Ca(OH)_2$ and mix it with the HCl solution, and the mixture forms a new solution of $CaCl_2$ and water. This fully neutralized solution is safe and won't harm your skin. This kind of reaction releases energy, however, because the acid and the base are high in energy, while the products that form after they react are low in energy. The excess energy is given off as heat. This is called an ***exothermic*** reaction (as opposed to an ***endothermic*** reaction, which requires heat to react). Some reactions give off so much energy that they become safety hazards.

It takes some time for all the molecules to find each other in the mixture and finish reacting. If enough heat is given off, the acid and/or base can actually boil or splash back when the solutions are first mixed together, since so much heat is given off. This is why very reactive chemicals must always be combined slowly in the laboratory. Many reactions take some time to complete (sometimes a matter of days).

In fact, all of the reacting molecules often do not find each other in the solution. When a reaction is performed in the laboratory, it is routine to calculate the expected amount of product and then, after the reaction is complete, calculate the percentage yield. This is done by dividing the actual weight by the expected weight and multiplying by 100%. For this example, our calculation would show that we expect 55.5 grams of $CaCl_2$ (the molecular weight of $CaCl_2$ is 111 g/mole). If in reality we get 35 grams of $CaCl_2$, we would calculate the percentage yield as follows:

(35g $CaCl_2$ actual/55.5g $CaCl_2$ expected) x 100% = **63% yield**

This is a very common calculation that you will encounter many times in your laboratory exercises.

Chapter Twelve—Review Questions

1. Why is a strong acid like HCl rendered safe after reacting with a base like NaOH or $Ca(OH)_2$?

_____

_____

_____

_____

2. Is heat given off by this reaction? Why or why not?

_____

_____

_____

_____

3. How many OH^- ions are in 0.5 mole of $Ca(OH)_2$?

_____

_____

_____

_____

4. Is it sometimes dangerous to mix acids and bases together? Why?

_____

_____

_____

_____

Afterword

I hope this book is helpful to you if you will be studying chemistry this year. I also hope that you are able to enjoy the fun parts of chemistry without having to struggle too hard to keep up, since often classes can go faster than you can absorb concepts. If you get behind in chemistry, it gets harder and harder to catch up, since each concept builds on the earlier concepts. I encourage you to work on as many of your assigned problems as you have time for, and get used to getting things wrong in class and at home, so you don't get them wrong on the test. Having trouble with this subject doesn't mean you can't learn; chemistry is just very complex and it takes time to absorb. Good luck!

HOW TO CONTACT THE AUTHOR

I would really like feedback on whether you liked or disliked the book. Please either post comments on the Web site if you purchased it online, send suggestions directly via e-mail to comments@tuxedopublishing.com, or visit the "Contact" page of www.tuxedopublishing.com and make sure you specify the title of the book. Thank you!

Appendix

FURTHER RECOMMENDED READING:

Once you've finished reading this book, you may be ready to move on to some problems that teach you more details about chemistry. One of these books may be helpful:

1. *Chemistry: Concepts and Problems: A Self-Teaching Guide,* by Clifford Houk.

2. *Chemistry for Dummies,* by John T. Moore.

3. *Basic Concepts of Chemistry,* by Alan Sherman, Sharon Sherman, and Leonard Russikoff.

4. *Chemistry the Easy Way,* by Joseph Maschetta (published by Barron's).

5. *Asimov on Chemistry,* by Isaac Asimov. This book is out of print but can still sometimes be found in libraries. It is a collection of essays about chemistry that are written in an easy-to-understand way and that give more insight into the historical context of the chemistry concepts you have read about in this book.

Alternatively, you might want to take a class. Here is one suggestion: http://www.chemistrysurvival.com/

I can't vouch for the quality of these books or courses; I'm only presenting them as possible options.

Abbreviations

atm—atmosphere, a measurement of pressure; 1 atm is the atmospheric pressure at sea level

C—Celsius

cm—centimeter

cm³—cubic centimeter (equivalent to one mL)

CN⁻—cyanide ion

H⁺—positive hydrogen ion, also known as a proton

HCl—hydrogen chloride; when dissolved in water it forms hydrochloric acid

HCN—hydrogen cyanide

K—Kelvin

L—liter

mL—milliliter

n—number of moles

P—pressure

pH—a measure of the acidity of a solution; possibly stands for "power of hydrogen"

R—the gas constant, 8.314 J/(mol K)

STP—Standard temperature and pressure (273K and 1 atm pressure)

T—temperature

V—volume

Glossary

Acid—see **Brønsted–Lowry Acid** and **Lewis Acid**

Anion—a negatively charged ion

Atom—the smallest indivisible piece of a substance that still has the same properties of that substance.

Base—see **Brønsted–Lowry Base** and **Lewis Base**

Bohr Model—a model of the atom, now considered obsolete, that depicts electrons orbiting the nucleus like planets around the sun. It is still sometimes used to introduce students to atoms.

Bond—an attraction between atoms.

Brønsted–Lowry Acid—a substance that ionizes in solution to produce more protons (H^+ ions) than can be found in neutral water.

Brønsted–Lowry Base—a substance that ionizes in solution to form more hydroxide ions that would be found in neutral water. It can also be thought of as a substance that accepts hydrogen cations (i.e., protons).

Calcium Hydroxide—$Ca(OH)_2$, a base (white powder). Used chiefly in mortars, plasters and cements.

Calorie—a unit of heat equal to 4.1840 Joules.

Cation—a positively charged ion.

Chemistry—the study of the composition and properties of substances.

Covalent Bond—a bond with one or more pairs of shared electrons.

Dipole—a situation where the electrons shared between two atoms move more toward one of the atoms, causing a slightly positive charge on the atom with fewer electrons and a slightly negative charge on the atom that ends up with more electrons.

Electrolyte—a substance that dissociates into ions when dissolved, forming a solution that conducts electricity. Electrolyte solutions are important in the body for regulating the flow of water between cells.

Electron—negatively charged subatomic particle.

Endothermic—a chemical reaction that absorbs heat from the environment.

Exothermic—a chemical reaction that gives off heat to the environment.

Hydrochloric Acid—HCl, a strong acid used industrially for ore reduction, food processing and metal cleaning. It is present in the stomach in small amounts.

Hydrophilic—water-loving; readily absorbing or dissolving in water.

Hydrophobic—water-fearing; having little or no affinity for or solubility in water.

Hydroxide Ion—OH^-; the ion that makes a solution basic.

Ideal Gas Law—the law that the product of the pressure and volume of one mole of an ideal gas is equal to the product of the absolute temperature of the gas and the universal gas constant, represented by the equation $PV = nRT$.

Ionic Bond—a bond between two ions of opposite charges, formed through the transfer of one or more electrons

Joule—a unit of heat equal to 0.239 Calories.

Kelvin Scale—the scale used in scientific work to measure temperature. The size of a degree Kelvin (K) is the same as a degree Celsius (C), but the Kelvin scale starts at absolute zero instead of the freezing point of water. The freezing point of water is equal to 273K.

Lewis Acid—a molecule that can accept an electron pair.

Lewis Base—a molecule that can donate an electron pair.

Logarithm—the power to which a base number, such as 10, must be raised to equal a given number. In chemistry, base 10 is most often used. Example: 3 is the logarithm of 1000 to the base 10 ($3 = \log_{10} 10^3$), often just shown as $3 = \log 10^3$.

Macro—very large in scale.

Metallic Bond—the type of chemical bond between atoms in a metallic element, in which potentially mobile valence electrons are shared among atoms in a crystalline structure. The movement of these valence electrons through a wire creates electricity.

Micro—extremely tiny in scale.

Model—a depiction to show the appearance of something not normally visible to the naked eye. Example: a model of an atom.

Mole—the mass in grams of 6.022×10^{23} atoms, molecules, ions, or other elementary units of a substance.

Molecule—two or more atoms bonded together very strongly to form a substance with consistent properties.

n—variable representing the number of moles of a gas in the Ideal Gas Law equation.

Neutron—uncharged subatomic particle.

Noble Gas—any one of the chemically inert gases found on the far right column of the periodic table (i.e., helium (He), neon (Ne), etc.)

Nucleus—the positively charged part of an atom located in its center, composed of protons and neutrons. Most of the mass of an atom is in its nucleus, but the nucleus only comprises a small fraction of the atom's total volume.

Octet Rule—the tendency of atoms to want eight electrons in their outer shell (valence shell).

Periodic Table—a table that organizes all the known elements so that their common properties can be more easily seen.

pH Scale—a scale that expresses the concentration of hydrogen cations (H^+) in a simple format. A neutral solution has a pH of 7, a pH reading of less than 7 indicates acidity; greater than 7 indicates basicity.

Pressure (P) —force per unit area, exerted against a surface.

Proton—positively charged subatomic particle.

R—gas constant, 8.314 J/(mol K).

Significant Digits (sig digs)—see Significant Figures.

Significant Figures (sig figs)—all the digits of a number that indicate accuracy; the number of sig figs is dependent on the accuracy of the measurement method used.

Sodium Bicarbonate—$NaHCO_3$, also known as baking soda.

Sodium Hydroxide—NaOH, also known as lye, used in the manufacture of paper and soap.

Solubility—the ability of a solid, liquid, or gas to dissolve in a solvent and form a solution.

Solute—a substance dissolved in another substance, forming a solution. Sugar that is dissolved in water is an example of a solute.

Solvent—a substance, usually a liquid, capable of dissolving another substance. Water, alcohol and benzene are examples.

Stoichiometry—the calculation of the quantities of chemical elements or compounds involved in reactions.

Strong Acid—an acid that dissociates completely in water to H^+ and an anion, giving it an extremely low pH.

Subatomic particle—a particle that makes up an atom.

Symbiosis—the association of two living organisms in a relationship that often benefits both of them (e.g. algae and fungus forming lichen).

Temperature (T)—a measurement of heat, almost always shown in degrees Kelvin (K) when working on chemistry problems.

Units—labels attached to numbers to indicate amounts (e.g., meters, grams).

Valence Shell—the outer group of electrons on an atom that is able to combine with the electrons in the valence shells of other atoms.

Vinegar—a dilute (3%) solution of acetic acid.

Volume (V)—the amount of space taken up by a gas, liquid, or solid.

Bibliography

Atkins, P. and L. Jones. *Chemistry: Molecules, Matter, and Change.* 3rd Ed. New York: W.H. Freeman and Company, 1997.

Carey, F.A. and R. C. Atkins. *Organic Chemistry.* 2nd Ed. New York: McGraw Hill, Inc., 1992.

Dictionary.com Unabridged. Source location: Random House, Inc., http://www.dictionary.reference.com. Accessed: November 15, 2009.

About the Author

Suzanne Lahl graduated with a B.A. in biology from the University of Pennsylvania in 1990. After working as a lab technician in several environmental and pharmaceutical labs, she returned to graduate school full-time and earned her Ph.D. in Organic Chemistry in 2002. While attending graduate school, she taught undergraduate students in introductory chemistry laboratory and recitations, as well as organic chemistry laboratory. She also served as a National Science Foundation (NSF) Teaching Fellow for one year, working with middle school teachers to help them infuse new science and math ideas into the classroom. For the past three years, she has been involved in volunteer work mentoring pre-teens and teenagers. She currently resides in Virginia and is working on a new book of educational advice for high school students. She can be reached through the website www.tuxedopublishing.com.

Index

Acidity IX, 49-51, 75-78

Aluminum foil 9-10

Anions 71

Antiperspirant VII

Atom 7, 8

Atomic number 15

Atomic weight 15

Basicity IX, 49-51, 75-78

Bonding X, 65, 67-72

   Covalent 70

   Hydrogen 71

   Ionic 68

   Metallic 71

Boyle, Robert 29

Calculator 23

Calories 68

Cations 71

Chemical symbol 15

Chemistry 1-2, 7

Chemistry problems 1

Conjugate acid 50

Conjugate base 50

Conversions IX, 42

Covalent bond 70

Cyanide 71

Dry cleaning 61

Electron dot structures 66

Electron pair 50

Endothermic 77

Exothermic 77

Greeks 9-10

Group 15

Homework 1

Hydrogen bond 71

Hydrophilic 62

Hydrophobic 61

Ideal gases IX, 53

Ionic bond 68

Joules 67

Lavoisier, Antoine 30

Mass spectrometry 11

Metallic bond 71

Micro 8

Models 10

Molarity 37

Molecule 8

Moles IX, 35, 45, 57, 75-76

Noble gas 69

Octet rule 66

Percentage yield 78
Period 15
Periodic Table IX, 7, 14
pH 51
Plastics VII
Pressure 55
Problems 1
Proportionality 56
   Direct 57
   Inverse 56
Proton 49-51

Qualitative properties 30
Quantitative properties 30

Reactions, chemical X, 75
Rounding 32
Rulers 30-31

Scale X, 10

Scientific Notation IX, 19
Significant digits
   See Significant figures
Significant figures IX, 29-32
   Addition 31
   Division 32
   Multiplication 32
   Subtraction 31
Soap 62-63
Solubility 61
Solute 37
Solvent 37
STP 37
System 53

Temperature 57

Units IX, 41

Valence electrons 66-67
Volume 55
VSEPR 67